Rearing Pheasants in Small Enclosures

by S.V. Reeves

with an introduction by Jackson Chambers

Self Reliance Books

Get more historic titles on animal and stock breeding, gardening and old fashioned skills by visiting us at:

Introduction

I am pleased to present yet another title on Raising Quail.

This volume is entitled "Quailology" and was published by Fred Kerr in 1903.

The work is in the Public Domain and is re-printed here in accordance with Federal Laws.

As with all reprinted books of this age that are intended to perfectly reproduce the original edition, considerable pains and effort had to be undertaken to correct fading and sometimes outright damage to existing proofs of this title. At times, this task is quite monumental, requiring an almost total "rebuilding" of some pages from digital proofs of multiple copies. Despite this, imperfections still sometimes exist in the final proof and may detract from the visual appearance of the text.

I hope you enjoy reading this book as much as I enjoyed making it available to readers again.

Jackson Chambers

QUAILOLOGY

Texan Bob White
Photo from life by Dr. R. W. Shufeldt.
(12a)

Chestnut-bellied Scaled Partridge

Photo from life by
Dr. R. W. Shufeldt

(18a)

(16a)   **Mountain Partridge**   American Ornithology.

Masked Bob White
Photo by Dr. R. W. Shufeldt from Seaton's Colored Plates
(14a)

PENS

LOCATE all pens or yards on the highest ground available, so as to facilitate proper drainage.

It is a good plan to provide shade by locating the yards, as much as possible, under fruit trees. If there are no trees that can be so utilized, then plant some kind of running vines.

These pens should be constructed of ¾ inch or 1 inch mesh galvanized poultry wire. The ¾-inch mesh wire is preferred, as 1-inch mesh will not exclude sparrows.

Where the grounds are limited in size the pens may be 8 x 16 feet, or even smaller. If smaller, the proportion indicated will be found advantageous.

In constructing these pens use ½-inch galvanized iron pipe, all uprights or posts being 8 feet long. Put them in the ground 2 feet or slightly less if desired, but have all of uniform height. Use a 12-inch wide base board all around the outer enclosure, placing the base boards on the outside of the posts. Before putting your pipe posts in the ground have one end, the upper, threaded for a fitting to carry such overhead piping as you may wish to employ. At same time as you have the threading done have a nail hole drilled through the pipe 2 feet 6 inches from the end that is put in the ground. This will bring the nail hole, when working for a 6-foot headway, opposite the center of the base board. Insert a wire nail and drive into board and clinch; this will keep pipes from settling down in the ground when soft from hard rains.

Should more head room be desired, say 6 feet 3 inches, this would allow the nail to pass through the base board 3 inches from top instead of in the center. In piping the top avoid long spans—have none greater than 4 feet without a support, unless you use ¾-inch pipe. With the latter size pipe the spans or unsupported sections may be as great as 5 or 6 feet. Where more than one pen is desired it is best to have them joined and have but one outside gate or entrance, as this will reduce the chances of birds escaping. In this connection, all gates should be provided with spiral springs to prevent the gates from being left open; an extra spring being used on the outside gates. Have all outside gates to open outwardly and all inner gates to swing both ways. Arrange a 6-inch wide removable board under all gates.

It is believed that all inner gates should be in the end of division fence farthest away from the shelter coops or houses.

After all posts and overhead pipings have been completed attach the wire netting to the top rail or pipe with annealed galvanized stove pipe wire. If the pens are to have a 6-foot head-way, use 60-inch wide wire, stapling the bottom of same to top of base board.

Dig a trench outside of base board about 7 inches deep by about 5 inches wide. Take four pieces of 1-inch mesh galvanized poultry wire 12 inches wide, and in lengths corresponding to the four dimensions of your pen or pens. With a straight edge of some kind form these strips of netting to a right angle, having one dimension 7 inches and the other 5 inches. Staple the edge of the 7-inch way to the outside of base board near bottom of same. This will carry the 5-inch way flat in the bottom of trench. Now fill trench and pack with a light broad surfaced rammer; or with the feet, as may be desired.

If you do not have gravel or sandy soil it will be necessary to procure same, spreading it over the natural soil to a depth of about two or more inches. Where it becomes necessary to put in sand and gravel it is advisable to put in a liberal coating twice a year—spring and fall.

During severe winter weather, when it is wet and cold cover the ground with hay or straw two or three times, being governed by the character of the weather and its duration. Similar provision must be made in late summer and fall for young birds—see page 14.

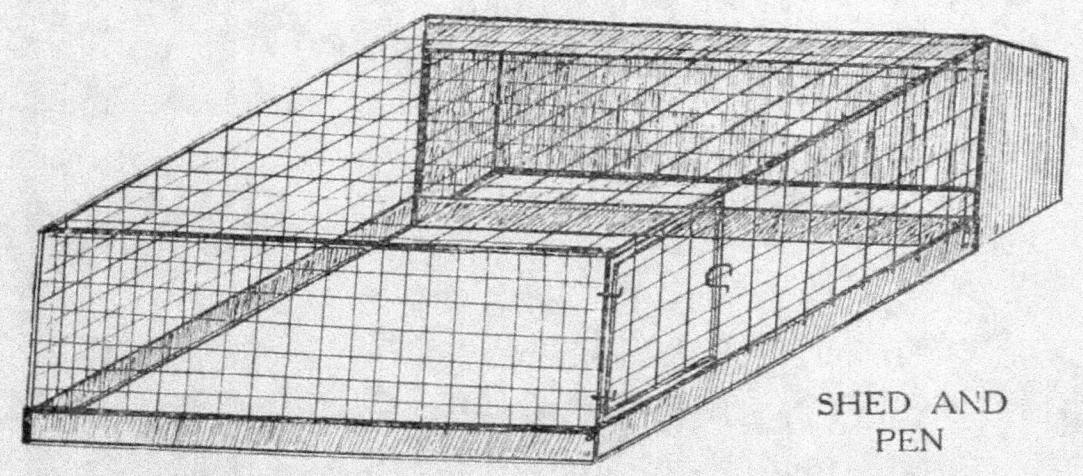

SHED AND PEN

SHELTER

All shelter sheds or houses should be quite open and located at one end of pen, and this should be the end farthest away from the gate entrance.

MATING

The number of hens that may be kept with a cock varies, but for best results the following will be found quite satisfactory.

Ring Neck, Prince of Wales, Mongolian, Versicolor, Reeves, Blue Neck, Lady Amherst and Golden each one (1) cock to three (3) hens. The following are mated in pairs: Silver, Swinhoe and Manchurian.

EGGS

In this locality pheasants begin to lay in the early part of April. Eggs should be collected frequently and stood on end in bran. Keep them in a dry place and turn about once a day.

Distribute a number of hard artificial eggs through your pens at the beginning of the laying season—it discourages egg eating.

FEED IN LAYING SEASON

Keep plenty of grit, oyster shells and granulated charcoal constantly before the birds. Do not feed corn in any form during the months of April, May, June and July. During these months feed oats, cabbages, apples, turnips, beef scraps, ground oats and bran, green clover, grass roots, lettuce and other greens. Occasionally a little wheat.

Feed green food of some kind frequently, but let it be in the morning as a rule.

HATCHING

Buff Cochin Bantams make splendid foster-mothers. They will cover from seven to nine eggs. Before setting the hen see that she is properly fumigated and free from lice. Do not put the eggs under the hen until you have tested her setting for twenty-four hours on a nest egg.

At the time of setting the hen dust her well with lice powder; also sift the powder well over the eggs. A separate enclosure is most desirable for setting hens. This may be a building or a yard, so arranged that the setting hens are separated, thereby preventing two or more hens occupying the same nest.

The nest should be in a hollow of the ground. Take a box or small barrel and cut a hole in the side near the top about 6 by 10 inches and turn upside down over nest. Six to twelve hours before the hatch is due cover this hole with a piece of cellar window wire to prevent the young pheasants from escaping.

During incubation feed mostly whole and cracked corn and keep the hens well supplied with sand, ground bone and shells; also charcoal.

If the hen sticks well to her nest for the first week you may count on the hatch being on time or perhaps a little ahead of time.

However, if she does not, the hatch will be a little late. Sometimes the variation from true time of incubation will be three or four days early or three or four days late. Where the hatch is quite uniform do not remove the young birds or hen from nest until about thirty hours from the time first one is hatched.

TIME OF INCUBATION

Ring Neck, Blue Neck, Lady Amherst, Versicolor and some others hatch in 24 days. Reeves, Silver Swinhoe and some others, 26 days. Goldens 21 to 23 days.

COOPS AND RUNS

My coops are made of half-inch thick planed lumber. Herewith is presented line illustrations of various parts of coop and run.

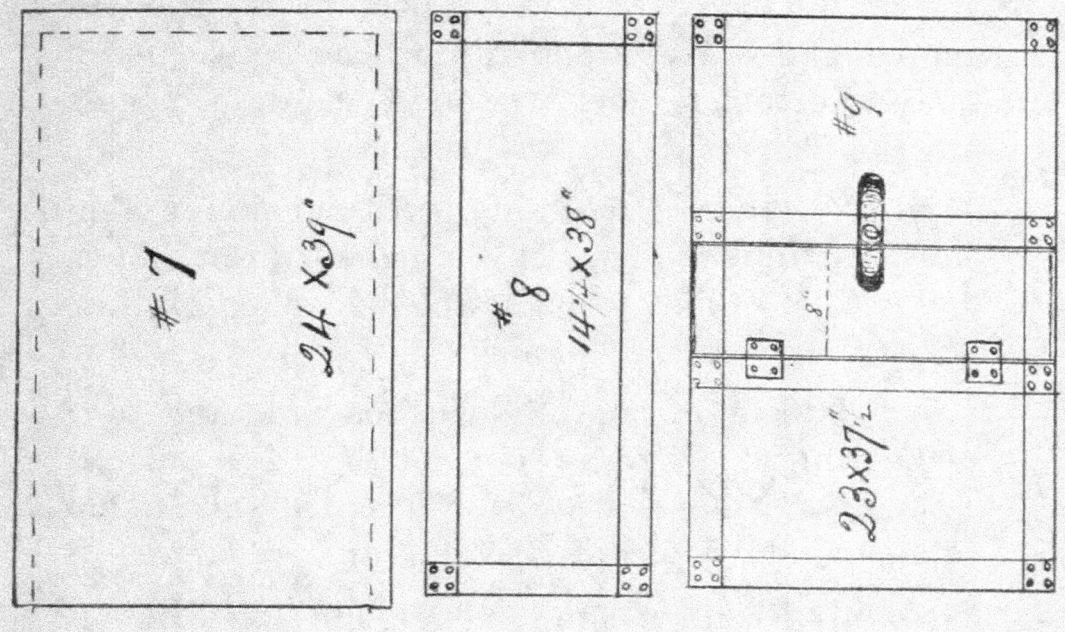

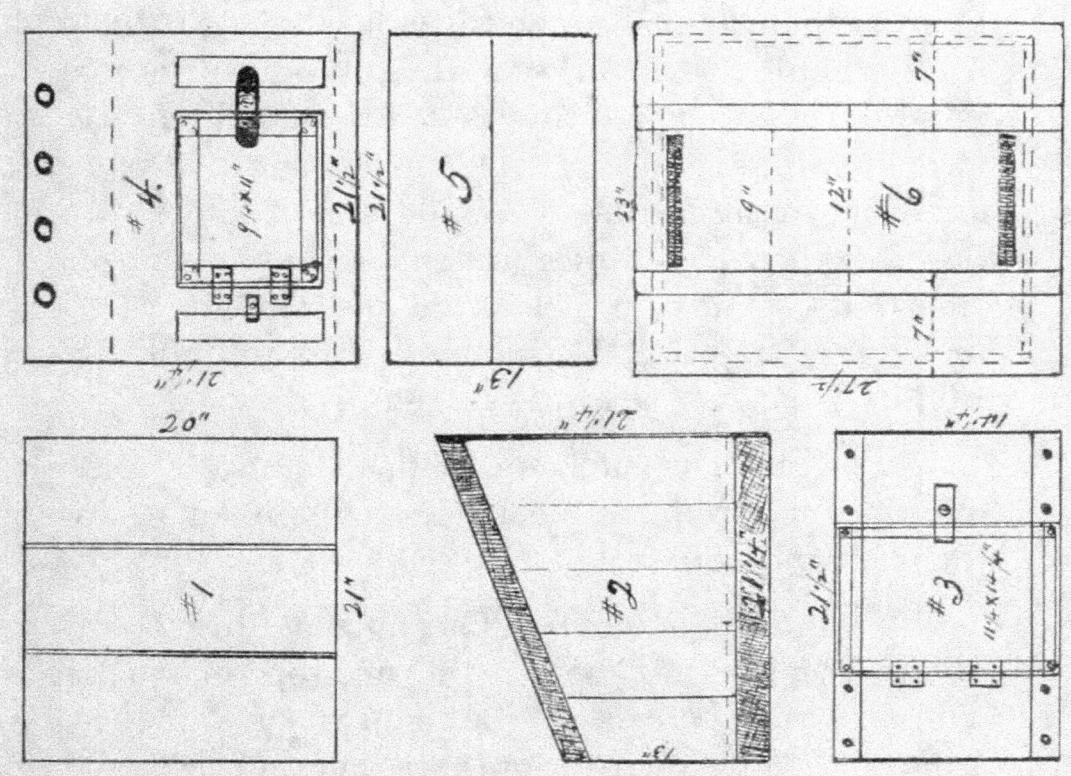

No. 1—Three (3) loose bottom boards for coop. These are supported on the cross pieces at the bottom of the two sides, see dotted lines on No 2.

No. 2—This is one side with the inner side shown; dimensions, 21¼ inches high, 21¼ inches from front to back and 13 inches high at back.

No. 3—End of run, 21½ inches wide and 14¼ inches high, with a door 11¼ x 14½ inches. This door should be covered with galvanized cellar window wire. Use galvanized hinges.

No. 4—Front of coop. The four holes in top are for ventilating purposes. The dotted lines near the top represent approximately top line of run when in position, and the dotted lines at bottom represent the floor line of run. The door, which is 9¼ inches wide by 11 inches high, should be covered with galvanized cellar window wire. On either side of this door are two openings about 2 x 11 inches for the passage of young birds. Provide a suitable stopper for these openings and hold them in position with small iron buttons as shown. Dimensions: 21¼ x 21½ inches. Use galvanized hinges wherever a hinge is required.

No. 5—This is back of coop. It may be one or more boards, as desired. Dimensions: 13 inches from top to bottom, 21½ inches long.

No. 6—Roof of coop. The dotted lines show about where the roof rests on top of coop. The two side boards of roof shown as 7 inches wide by 27½ inches long are nailed fast to top of coop. The board in center shown as

12 inches wide has a cross piece scant 9 inches long on under side nailed across each end, as shown, to keep it in place and to keep it from warping. This board is loose and covers a 9-inch opening in top of coop, overlapping the 7-inch wide boards 1½ inches on either side. Dimensions: 23 inches wide by 27½ inches long.

No. 7—Bottom board upon which the run rests. The under side of this should have two cross pieces about 2 inches thick. Dotted lines show position of run. Dimensions: 24 inches wide by 39 inches long.

No. 8—Side of run. Cover these with galvanized cellar window wire. Dimensions: 14¼ inches high by 38 inches long.

No. 9—Top of run. Door can be located in center or near one end as may be desired. This lid or door is made of wood. It is about 8 inches wide and corresponds in length to full width of top.

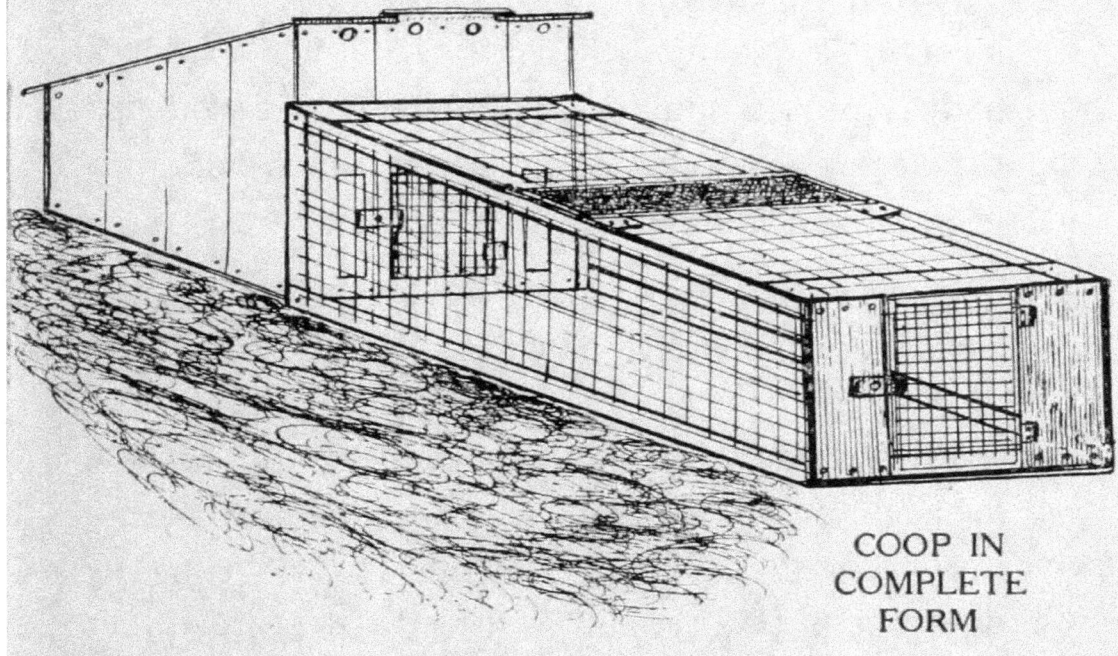

COOP IN
COMPLETE
FORM

Before taking the hen off paint the coop well inside with a mixture of cresol and coal oil. Have the proportions about half pint cresol to one gallon of coal oil.

It is a good plan to keep young birds in with the mother for two to four days—until they have become accustomed to her call. After this the young birds should be let out into the run through one or both of the openings at side of door.

Cover bottom of coop and run with clean sharp sand, adding a little granulated charcoal.

It is a good plan to provide oil cloth or other waterproof covers for these runs, to be used during rainy weather and nights.

Examine the hen and young pheasants at least once every week for lice. I use an ointment, appling same to top of head and vent for the first five weeks, or longer if the weather is warm.

FEEDING YOUNG BIRDS

First 3 days—To 1 egg well beaten add 1 tablespoon full of milk, cook very slow, forming a dry custard. Feed often, say six times a day, but only as much as they will clean up.

Next 4 days—Feed as before, but add a little dry oatmeal and lettuce cut very fine; also bread crumbs.

Next two weeks—Feed as before, but add millet and canary seeds. Four times each day will be sufficient.

At end of 3 weeks—Commence now to feed three times each day. Add to above a very little meat, ground fine; also a little barley meal. Continue for two weeks.

When 5 weeks old—Feed as above, but in much smaller proportions. Add meal, greens, mashed potatoes, boiled rice, etc. At night feed wheat and oats or barley alternately.

Drinking Fountains—Use small fountains so the young birds can not get in them with their feet. Under no circumstances give water oftener than twice a day, and never allow the water to remain before them after they are through drinking.

CAUTION

Keep all vessels absolutely clean.

Remove droppings of hen from coop twice daily.

During spring and summer scrape yards once each week.

Renew sand in coops and runs once each week.

Sprinkle dry air slacked lime over ground frequently.

Keep birds dry—not exposed to dampness until at least half grown.

Keep ground bone, charcoal and various kinds of grit constantly before the young birds from the day they are taken from the nest.

Never attempt to rear young pheasants without lettuce. Feed it as often and as long as you can.

When the birds are about two months old cover the ground in their pens with plenty of leaves, straw, hay, etc., and provide plenty of roosting poles in the open.

NAMES OF A FEW WELL KNOWN VARIETIES OF PHEASANTS AND THEIR NATIVE HABITATION

Mongolian (P. Mongolicus), Valley of Syr—Darya, China.

Chinese Ring Neck (P. Torquatus), Northern and Eastern China.

Japanese (P. Versicolor), Japan.

Reeves (P. Reevesii), Mountains of Northern and Western China.

Prince of Wales (P. Principalis), Northwestern Afghanistan and Northeast Persia.

Great Argus (A. Giganteus), Malacca and Siam, Malay Peninsula, and Southern Tenasserim; also Sumatra.

Manchurian Eared (Crossoptilon Manchuricum), North of Pekin, Manchuria.

Swinhoe Kalij (Gennaeus Swinhoii), Mountains of Formosa.

Lady Amherst (Chrysolophus Amhersitae), Western China and Eastern Tibet.

Silver (Euplocamus Mycthemerus), Interior of Southern China.

Golden (Thaumalea picta), Western Central China.

Monaul or Impeyan (Lophophorus Impeyanus), Himalaya Mountains from Afghanistan to Sikhim.

Elliots (Calophasis Ellioti), Mountains of Southern China.

www.ingramcontent.com/pod-product-compliance
Lightning Source LLC
Chambersburg PA
CBHW081653220526
45468CB00009B/2629